WE'RE TALKING NOW!

USING VOICE IN YOUR BUSINESS

DAWN HARGROVE-AVERY

Voice Wizards

Voice Wizards
2396 Caton Crest Dr.
Crest Hill, IL 60403
www.VoiceWizards.com
Edition One: July 2020

ISBN 978-1-7354043-0-1 (paperback)

ISBN 978-1-7354043-1-8 (eBook)

Printed in the United States of America

EARLY ADVOCATES

Ann Hargrove

Patrick Sweetman

Sameer Rasa

David Winford

I wanted to take a moment to thank a few people, without them I would have never had the courage to complete this project.

First off, my mom who sometimes pushes me way out of my comfort zone, Thank you!

My Kids, Nicholas, and Jacob while you can be overly critical of my work at time, I appreciate your support, truthfulness, faith, and youthful thoughts. You two are truly the best thing I have ever done, love you always.

Patrick, in life I believe we cross paths for a reason, you are another one that has pushed me way past my comfort zone and shown me a whole new world in voice. Sameer, the man behind a lot of the magic, you are an amazing and highly intelligent person. I have enjoyed working with both you and Patrick over the past couple of years.

David, what started as two people working together to help a small business has turned into a lifelong friendship. I deeply appreciate you and everything you do. Our conversations are exceptionally long 😊 but worth every single minute. I hope you feel the same.

TABLE OF CONTENTS

INTRODUCTION

Let me start by sharing some of my background with you. It is that journey that has led me to Voice, an unanticipated destination.

I grew up in the dry-cleaning industry, working in my mother's suburban Chicago dry cleaner business. The work was challenging. A dry cleaner has a tropical environment all its own – minus the beach chairs and sand. In addition to the excess heat and hard work, as the owner's daughter, I was expected to jump into multiple roles as needed – sick employees, late employees, broken equipment, etc. Jobs had to be completed; clients depended on us. In any business, customer service can be like a box of assorted chocolates – when speaking with a client you could be dealing with anger, disappointment, or gratitude.

After getting married and starting a family, I decided dry cleaning was not for me and pursued a career in a hospital. Because the hospital offered tuition reimbursement, I chose to go back to school: my major, computer science. While attending classes, I learned a variety of programming languages and business applications.

When I finished, I had quite a dilemma. The thought of sitting in a cubicle five days a week, 8 hours a day with minimal human contact did not appeal to me. While I am a self-proclaimed introvert or "shy," as I like to call myself, I probably fall somewhere between introvert and extrovert—just an average person who likes to spend a lot of time alone but also welcomes human interaction.

The tech world and the internet had arrived, and businesses were looking at the power and efficiency of the internet as a business tool. With my programming background, I quickly taught myself HTML, Java, Cold Fusion, and a few other graphics programs. My goal was to teach small businesses how to utilize the internet to promote their businesses.

I accepted an opportunity to become the Internet Sales Manager at a car dealership. I then went on to a landscaping company close to home. It was a startup company that provided a directory to landscapers on the internet.

Around this same time, my mom sold her dry-cleaning operation and was working with a company to promote an alternative solvent for the dry-cleaning industry. While I was helping her, we became acquainted with several small business owners. As I became friends with this group of people, I realized their problems were similar—training, marketing, acquiring new customers, adhering to government regulations and, of course, using new technologies.

For the past 15 years, I have focused on small businesses, primarily service-related companies, and their internet presence. From websites and SEO to digital marketing and strategies, I have also worked with print advertising. I provide clients with the resources, tactical know-how, and a full suite of tools to assist them in getting noticed and generating leads online.

Because I have always been a tech geek and intrigued by the latest tech gadgets, I found myself fascinated when in November 2016, Amazon marketed the Echo Dot smart speaker. I had no idea what it was, just that I thought I needed it and had to have it TODAY, and I believed everyone in my family also needed this fantastic little speaker.

The day my Echo Dot arrived, I immediately set it up, and I did what everyone at the time was doing. I played music, asked silly questions, researched recipes, checked the weather, and because I really love lists, I made countless lists. As I proceeded to interact with the gadget, I questioned everything— How does it know? Why did it tell me that? Where is the information coming from? Is the Dot always listening?

I then thought about how this technology could help small businesses. What if it could send emails, call clients, schedule appointments, or complete the tasks that your websites and apps are currently doing?

Late spring of 2018, there was a webinar for the dry-cleaning industry on disruptors, and in the past, we had talked about companies, beacons, the dash button, and service disruptors. During the webinar, all these things were mentioned, along with Alexa and the Echo Look. I was excited at the time because it validated my thoughts and my ideas about the what ifs. Someone else saw the same thing I was seeing. This was beneficial because now I had someone to share ideas with. One thing led to another, and we decided to build an Alexa skill for dry cleaners.

We wanted a skill that customers could use to find general information about the dry cleaner or schedule a pickup. In an effort to avoid lines or have a long wait, we wanted a skill that would notify the dry cleaner when the customer was on the way to pick up their order—creating somewhat of a curbside pickup or VIP program.

New questions arose that needed to be answered.

o How will people find you?

o What other technologies are out there?

o Is Google going to play nice with Amazon?

o What about Siri?

o Now there is Bixby, how does that fit into the picture?

o What about my new computer that came with Cortana? Who is she? What is she?

We had so many questions and did so much research. Sometimes the research led to more questions than answers. In an effort to help others on the same path that we were on, we recorded our findings in a document. I am hopeful that this will do a few things for you.

- Reduce your learning curve a bit.

- Assist small businesses with a reference manual and workbook to help you get heard and get your company on Voice.

- Help other digital marketers by giving them information that might enable them to provide clients with an add-on service using Voice technology.

We had the first Dry-Cleaning skill that went to market, which was exciting, but it was probably about 3 to 4 years too soon for the dry-cleaning industry. In this book, I am going to provide you with tips on finding a skill, where the information comes from, how to map out a skill, how to remove the frictions, what you should pay attention to, tips on getting ahead, and some future predictions.

I intend to provide you with some basic terminology that will help you answer these questions:

1. **What is Voice?**

_____
_____
_____
_____

2. **What is a Voice skill or action?**

_____
_____
_____
_____

3. **Are you prepared for Voice?**

_____
_____
_____
_____

4. **How can you prepare for Voice?**

_____
_____
_____
_____

ARE YOU READY FOR VOICE?

What is Voice?

Are you ready for Voice? What the heck is Voice? What does that mean? Do you have a Voice skill or action? These questions are somewhat new to the general public as well as many small business owners. It became apparent while speaking with thousands of small business owners over the past few years. This book can help regardless of your familiarity with Voice, Voice technologies, skills, and actions, Alexa, or Google.

To understand, we need to start at the beginning by providing you with information that will assist you as a business owner or marketing professional. Let us spend some time defining Voice terminology, giving you a better understanding of what Voice is, what Artificial Intelligence (AI) is as well as the devices and platforms currently in the market.

Voice Technology Basic Definitions

Voice: Using natural language to have an interaction with any computer-like device. Desktop, laptop, mobile phone, smart speaker, screened device, wearable, and build in devices.

Current Devices (as of April 2020)

Smart Speaker: A smart speaker is a wireless, Voice-activated device that uses integrated virtual assistant software to obtain information or perform tasks and provides results that the user can hear.

Smart Display: A wireless, Voice-activated device that uses integrated virtual assistant software to obtain information or perform tasks and provide the results for the user to hear as well as present visual information on a screen.

Smart Phone: A mobile phone with highly advanced features. A typical smartphone includes a high-resolution touch screen display, Wi-Fi connectivity, Web browsing capabilities, and the ability to accept sophisticated applications. The majority of these devices run on any of these popular mobile operating systems: Android, Symbian, iOS, BlackBerry OS, and Windows Mobile.

Wearables: Portable smart devices that are worn on the body. Wearables include devices such as smart glasses, i.e. Google Glasses, and smart jewelry, for example Ringly.

Hearables/Smart headphones: Technically advanced, electronic in-ear devices designed for multiple purposes ranging from wireless transmission to communication objectives, medical monitoring, and fitness tracking.

In-Car: Speech recognition systems devised to eliminate the distraction of looking down at your mobile phone while you drive. Instead, a heads-up display allows drivers to keep their eyes on the road and their minds on safety. Companies, such as Apple, Google, and Nuance are reshaping the way Voice-activation is used in vehicles (https://www.globalme.net).

In-Appliances: When Voice assistants are used with consumer devices, simple single-action commands are the most popular. For instance, changing the temperature of a thermostat or starting an oven.

Facebook- Portal (powered by Alexa): Facebook Portal is a brand of smart displays developed in 2018 by Facebook, Inc. The devices are integrated with Amazon's Voice-controlled intelligent personal assistant service Alexa and work similarly to a screened version of Alexa.

Companies and Brands as of 2020

Alexa: Alexa is Amazon's cloud-based Voice service available on more than 100 million devices from Amazon and third-party device manufacturers. Alexa affords you the opportunity to build natural Voice experiences that offer customers a more intuitive way to interact with the technology they use every day (https://developer.amazon.com). In addition, Amazon offers a collection of user-friendly tools, APIs, reference solutions, and documentation to make it easier to build skills for Alexa.

Google: Google Assistant is an artificial intelligence-powered virtual assistant developed by Google that is primarily available for mobile and smart home devices. Google offers many tools making it accessible to develop and build actions for Google devices.

Apple-Siri: Siri is a personal assistant that resides on your iPhone 4S (and future iPhone generations). Siri responds to the words you speak rather than requests you type. Just as you can talk to your iPhone 4S to perform a range of tasks (employing speech-to-text translation), you'll also hear Siri's human-like Voice speak back to you (text-to-speech technology).

Samsung-Bixby: Bixby is the Samsung intelligence assistant first introduced on the Galaxy S8 and S8+. You can interact with Bixby using your Voice, text, or taps. It is deeply integrated into the phone, meaning that Bixby is able to carry out many of the tasks you do on your phone.

Microsoft-Cortana: Microsoft's virtual assistant for Windows desktops and smartphones. Introduced in 2014 for Windows Phone 8.1, Cortana was added to Windows 10 in 2015. Side-by-side Voice queries showed that Cortana provided similar answers to Siri and Google in most cases. Cortana includes a predictive notification function similar to Google, and it offers event-based notifications that let you request, for example, "When your brother calls, tell him congratulations." Because Skype is a Microsoft product, Cortana provides smooth activation of Skype video calls from Voice commands. Cortana is also integrated into Microsoft Edge, the replacement for Internet Explorer.

Connections by interactions

Voice Only: An application in which Voice is the only input and the only output. True Voice-only interactions are scarce. Even those situations that place the heaviest emphasis on Voice often have an alternate mode. At present, the Amazon Echo and Google Home are the most recognizable applications.

Voice First: When an interaction relies primarily upon, but not exclusively, on Voice for the input and output. Voice-first is generally characterized by screens as a nearly equal output, which can sometimes be used as input. Common Voice-first devices include the Echo Show and Spot, plus televisions and set-top boxes with Voice input.

A device can switch from Voice-only to Voice-first or Voice-added to Voice-first. Voice-only becomes Voice-first with the addition of new modes, such as connecting Echo Buttons to an original Amazon Echo.

Voice-added becomes Voice-first with the loss of modes, like a Pixel phone placed on a dashboard.

Voice Added: When Voice is no longer the primary tool for input or output. Voice-added is usually seen on mobile, where Voice is added as an additional input mechanism to alleviate the inconvenience of typing on an on-screen keyboard.

General Terms

Digital Strategy: Focuses on using technology to improve business performance, whether creating new products or reimagining current processes. Digital strategy specifies the direction an organization will take to create new competitive advantages with technology, as well as the tactics it will use to achieve these advantages.

Digital Marketing: The marketing of products or services using digital technologies on the Internet, through mobile phone applications, display advertising, and any other digital media.

Long-Tail Keyword: are keywords and key phrases that are specific and typically longer then regular keywords.

Schema markup: is code that you add to your website that helps the search engines return informative results to the user. It tells the search engines what your data means. Like HTML that tells the browser how to display your data. You can learn more at Schema.org

SEO: Stands for "search engine optimization." It is the process of getting traffic from the "free," "organic," "editorial," or "natural" search results on search engines.

Third-Party App: An application that is provided by a vendor other than the manufacturer of the device. For example, the iPhone comes with its own camera application, but there have been camera applications from third parties that offered advanced features such as a self-timer and simple editing.

UI: (User Interface Design) is the design, look and feel of a company's services and products. Very strongly associated to UX.

UX: (User Experience) all aspects of an end user's interactions with a company, it is services and products.

Voice Marketing: The use of marketing strategies and tactics to reach your target audience through the use of Voice-enabled digital devices.

Voice Strategy: The actions implemented in your overall marketing strategy that include your company's value proposition, essential branding messaging, customer demographics data, and other elements of your plan. It specifies the direction an organization will take to create new competitive advantages with technology, as well as the tactics it will use to achieve these advantages.

Voice Search Optimization: The output or results for typed vs. spoken searches will deliver different results. This means that typical search engine optimization will not work the same as it has in the past. One of the largest concerns regarding Voice search is that when Voice search is used on mobile you may only get one top result. Optimizing for Voice search is going to be incredibly significant in the years to come.

Now that you understand the terminology, let us dig a bit deeper.

WHERE IS ALL THIS INFORMATION COMING FROM?

Voice assistants provide an overwhelming amount of information to consumers and businesses alike. *Where does all this information come from? How is it found? What are the key factors in the results delivered?*

Amazon's Alexa Informational Sources

When you ask Alexa a question, it is gathering general information from Bing and pulling local information from Yelp. As a business owner or marketing agency concerned with optimizing Alexa, also be sure to optimize Bing places, Yelp, and Google's featured snippet area. The Alexa skill store is available for consumers and businesses to seek out entertainment and leisure skills.

Google Assistant Informational Sources

When using Google Assistant, general information is coming from Google Search and the featured snippet area. Local information is gathered from Google My Business, formerly called Google Local and Google Places. It is where your company's business information is listed on Google. The Assistant Apps store via Google Play is available for consumers and businesses to seek out entertainment and leisure skills.

Samsung's Bixby Informational Sources

Bixby uses Wikipedia and Google to gather information to answer your questions. Bixby also queries Bixby Capsules, their marketplace store. This is a little different than Alexa and Google at the present time. The Bixby Marketplace is available for consumers and businesses to seek out entertainment and leisure skills.

Apple's Siri Informational Sources

Siri uses Google to search for general information and Google My Business for local information. Siri on Home Pod will search iTunes and the App store for consumers and businesses to seek out entertainment and applications.

Microsoft's Cortana Informational Sources

Cortana currently gathers both general information and local information from Bing. This is probably no surprise considering it is Microsoft. For entertainment and leisure, there is more in the Cortana skill store.

More on Applications, Skills, and Actions

There are also third-party skills, actions, and capsules available for use in the marketplaces of Amazon, Google, Samsung, Siri, and Cortana. Third-party applications are created by companies and brands to provide a place that consumers or clients can go to interact with the company using Voice.

- Alexa refers to these third-party applications as *Skills*.

- Google calls them *Actions*.

- Bixby calls them *Capsules*.

- Siri is calling them *Apps*.

- Cortana is calling them *Skills*.

These tools are similar to applications on a smartphone. There are many 3rd party-generated skills on the market, and the field is growing rapidly.

DEFINING VOICE AND ARTIFICIAL INTELLIGENCE (AI)

If you were to break Voice down to its simplest and most basic definition:

Voice is to express something in words.

AI is the simulation of human intelligence processes by machines, especially computer systems.

Voice is expressing something in words or having a conversation, and AI is simulating these conversations using a computer system.

Conversational AI is software and technologies that allow computers to simulate authentic conversations.

Voice Assistant is a digital assistant that uses Voice recognition, natural language processing, and speech synthesis to provide aid or help users through devices, such as smartphones, speakers, and applications.

Voice assistants can interpret human speech and respond using synthesized Voices.

There are smartphone-based Voice assistants as well as smart speaker assistants and computer-based assistants.

As of the final quarter of 2019, assistants in the marketplace consist of the following:

Smart Phone	Smart Speaker	Computers
Google Assistant	Alexa	Cortana
Alexa	Google	
Apple Siri	Portal (Facebook)	
Samsung Bixby	Apple Home Pod	

Alexa uses the wake word, "Alexa." You can make calls, play music, play games, locate businesses, set alarms, ask questions as well as provide consumers with third-party skills. You can also utilize Alexa to set up business processes in your business, companies, and workplace and create simple Voice command automation.

Google uses the wake word, "Hey Google." You can make calls, play music, play games, locate businesses, set alarms, ask questions as well as provide consumers with third-party skills. You can also utilize Google to set up business processes in your business, companies, and workplace and create simple Voice command automation.

Siri uses the wake word, "Siri." Siri is a virtual assistant that is part of Apple Inc.'s iOS, iPadOS, watchOS, macOS, and tvOS operating systems. The assistant uses voice queries and a natural-language user interface to answer questions, make recommendations, and perform actions.

Bixby uses the wake word, "Bixby." Bixby is a virtual assistant created by Samsung Electronics, that makes it easier to use your phone. Bixby learns, evolves, and adapts to what you like to do, working alongside your favorite apps and services to help you get more done. Bixby will remember how you interact with it, to give you a more individualized experience. The more you use Bixby, the better it will become at adjusting to your needs. You can build Bixby capsules similar to an Alexa Skills or Google Actions.

WHO ARE THE CONSUMERS USING THESE DEVICES?

In order to understand who is using these devices from a business point of view, let us review what is currently available in the marketplace. Consumers were asked, "Which business types would you use Voice search for?" This is how they responded. (Rosie Murphy, 2018)

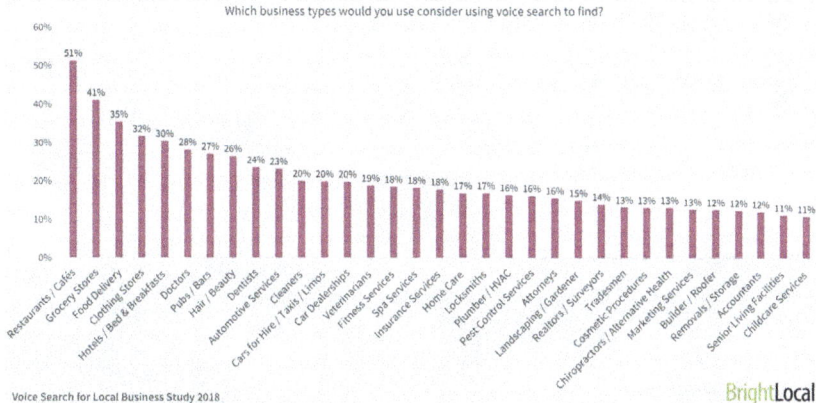

Which business types would you use consider using voice search to find?

Voice Search for Local Business Study 2018   BrightLocal

Which business types would you use Voice search for?

As you can see, consumers are using Voice to search all types of businesses, from restaurants to child care services, and many in between. Many small and large businesses are embracing this new technology and are considered the early adopters. They represent a variety of sectors, such as healthcare, insurance, finance, automobile, higher education, government, brands, media, travel as well as new fields such as Voice and AI.

> *According to Geo Marketing, "65 Percent of people who own an Amazon Echo or Google Home can't imagine going back to the days before they had a smart speaker."*

Another question that this group of consumers were asked was: Have you used voice search to find information on local businesses in the past twelve months ? (Rosie Murphy, 2018)

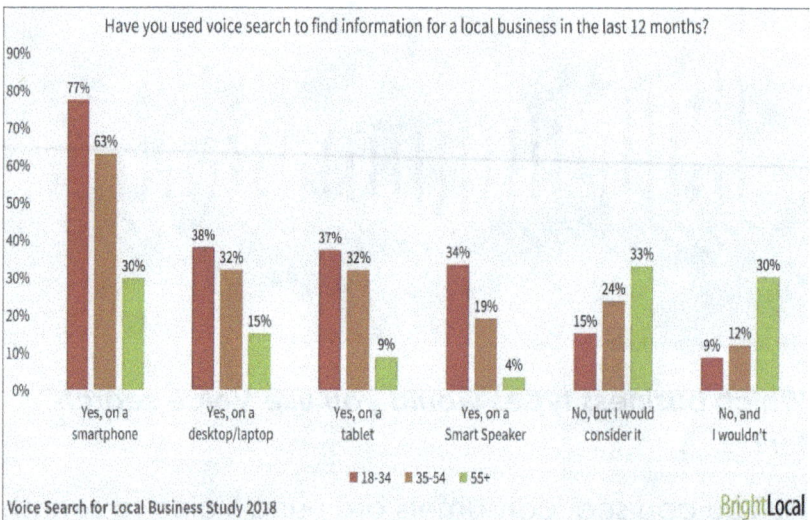

Have you used voice search to find information for a local business in the last 12 months?

	Yes, on a smartphone	Yes, on a desktop/laptop	Yes, on a tablet	Yes, on a Smart Speaker	No, but I would consider it	No, and I wouldn't
18-34	77%	38%	37%	34%	15%	9%
35-54	63%	32%	32%	19%	24%	12%
55+	30%	15%	9%	4%	33%	30%

Voice Search for Local Business Study 2018   BrightLocal

As you can see many have and the age group of 18-34 seemed to have the largest usage rate followed by the 35-54 year old, what was very encouraging to me was that 55 and up are getting familiar with this technology. Another thing to point out here is that most were using voice search on their smartphones in this 2018 study.

Frequency of use comes to mind when trying to determine if this technology is going to have loyal users. An additional question asked in the Bright Local Voice Search Study was how frequently do you use voice to search to find information on local businesses? (Rosie Murphy, 2018)

This question showed the smart speaker had the highest percentage of daily users followed by the smartphone. Overall, there was a fairly high percentage of people using voice search on a daily basis to find information about local businesses and this number is growing every day.

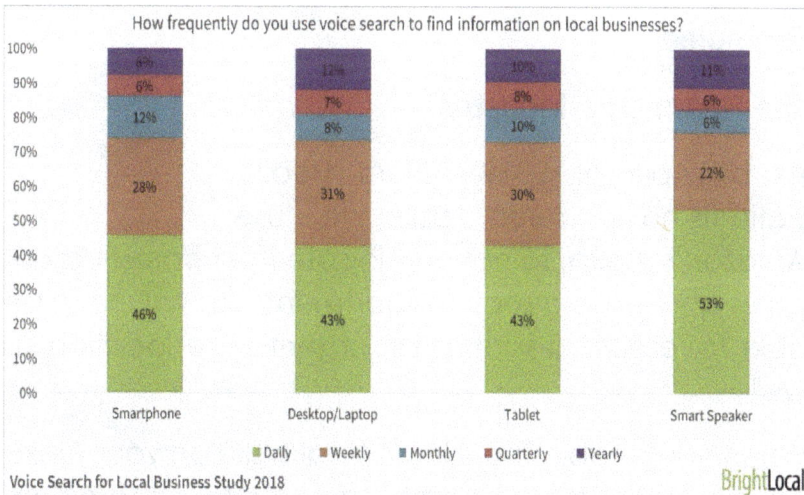

How frequently do you use voice search to find information on local businesses?

Voice Search for Local Business Study 2018

BrightLocal

31

WHICH BUSINESSES ARE INVESTING IN VOICE?

The healthcare industry is investing in Voice technology by allowing patients to look up Wait time, Waiting-List Signups, Video Calls, Patient-File Access. Initially, HIPPA laws hindered Voice-technology use; however, these challenges have been resolved as of April 2019 when Alexa became HIPAA compliant. See the following examples of the ways businesses and industries are using Voice skill technology.

Examples

The Healthcare Industry

As of April of 2019 Atrium Health patients have been able to use Amazon's Alexa, to not only locate the nearest urgent care and emergency department but will now be able to reserve a spot at an urgent care location as well.

When you enable the Atrium Health skill on your Alexa device, patients just need to say, **"Alexa, open Atrium Health."** Alexa will respond and guide a patient through the process of reserving a spot at one of Atrium Health's 31 urgent care locations. Patients can also get wait times at urgent care or emergency room facilities, as well as phone numbers or urgent care hours.

Providing this skill through Voice technology is part of Atrium Health's commitment to making healthcare easier to access and manage. Atrium Health is one of a few healthcare providers in the nation with Alexa "skills." (Atrium Health News, 2019)

Go ahead check it out Say: **"Alexa, open Atrium Health."**

The Insurance Industry

The Insurance industry has embraced Voice technology by allowing consumers to get a quote, fill out an insurance application, pay insurance premiums, and even file a claim using a skill.

Check it out here: "Alexa, open Geico."

GEICO is utilizing Voice Assistants to make it easier to connect and manage your policy 24/7. Things you can do with the GEICO Alexa Skill:

- Get updates on your claim
- Request roadside assistance
- Check your balance and next payment date
- Check your total premium
- Find a local agent

- Find gas near you
- Define Insurance terms

Banking and Finance

Financial Services are getting on board by giving their clients the ability to check balances, locate branches, pay credit card bills, look at market portfolios, and listen to market and finance podcasts.

> Try this one: "Alexa, open Capital One."

Ask Capital One about your credit card, checking, savings, and auto loan accounts. Then try out our new feature, "How much did I spend?" to get quick answers on how much and where you are spending.

All you have to do is enable the skill, say, "Alexa ask Capital One..." and you will be able to as a number of questions regarding your credit cards, auto loans and banking. You will be able to ask things like, "How much did I spend at Target last month?", · "How much did I spend at Starbucks last weekend?", "When is my credit card bill due?". You can also ask "What's my checking account balance?", "What are my recent transactions?".

Educational Institutions

Higher Education institutions are becoming Voice savvy by providing students the ability to check class schedules, final schedules, lecture, and lecture playbacks along with disability support.

All you have to do to try it is say, "Alexa, ask A.S.U when the next football game is"

You can use the skill to find out about ASU events to attend, business hours for popular ASU businesses and buildings, frequently asked academic calendar dates or random ASU facts. You can even ask Alexa to play the ASU fight song!

Government

The government is involved with Voice by providing road conditions, local events, and tourist information.

"Alexa, ask Mississippi a random fact."

The State of Mississippi has a skill on Alexa. With ms.gov's "Ask Mississippi" skill, Alexa can find information located on ms.gov quickly. All you have to do is ask. Once you enable the skill you can ask all kinds of things like when your driver's license expires (requires a myMS account) or how to renew your hunting and fishing license.

Big Brands

Big brands are getting on Voice with product information, DIY tutorials, recipes, and in-skill purchasing of products using order and reordering features.

> **Big brands like Clorox - "Alexa, open Clorox Clean."**

Clorox Simply checks in once a day, and Alexa will guide you through one house cleaning task with simple-to-follow steps. To dial up the fun, she'll also share all kinds of neat facts and handy tips along the way.

Media's involvement has gone beyond just listening to the radio or channel; it has spread to podcasts, ratings, and audiobook listening.

Entertainment

> **"Alexa, ask iHeart for country."**

With the iHeart skill, you can listen to over 2,000 radio stations nationwide, including iHeartRadio exclusives like The Classic Rock Channel and American Top 40. You can ask iHeart to play music from a specific genre or ask for your favorite local station by name.

There are also Alexa 4 Musicians (https://interactivealbum.app/) based out of Columbus, Ohio. Created by Voice First AI Alexa 4 Musicians is the only platform in the world for musicians to create experiences on Amazon Alexa for their fans. Currently, Voice First AI has the most musician skills published globally, and it is growing every day.

Travel Industry

The Travel industry is providing consumers with some useful features such as schedule confirmation, account lookup, ticket changes, loyalty status as well as controlling hotel rooms using Alexa to turn lights on, check out, and many other features.

Airlines are utilizing Alexa, you can say: "Alexa, ask United what's the status of flight 233."

The fairly new United skill gives Alexa the power to "ask United" for flight status information and what amenities are on your flight — and best of all, you can have Alexa check you in for your United flight if it's within the U.S. Your Voice and the simple command kicks it all off.

The above samples shown are just a few of the skills located in the Amazon Skill store. There are currently over 100,000 skills available in the Skill store that are in addition to Alexa's built-in features. Many large brands will continue to invest in Voice in the coming years.

RECENT VOICE USAGE AND IMPLEMENTATION STATISTICS

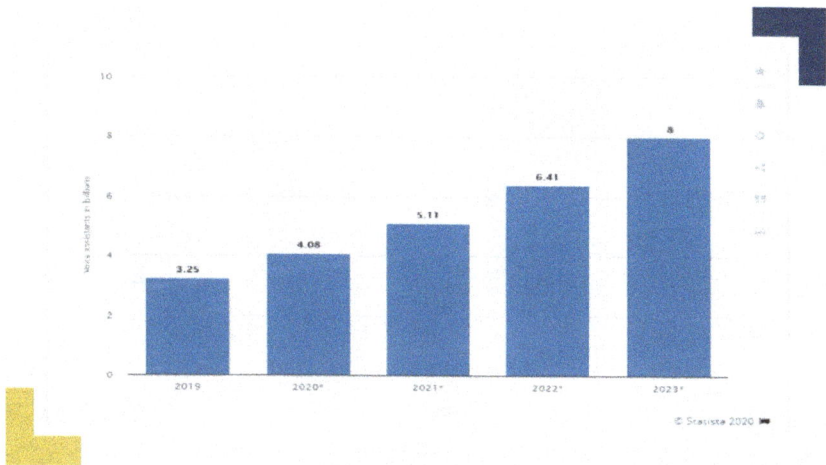

As of March 2020 *(Statista 2020)*

As of 2019, there are an estimated 3.25 billion digital Voice assistants used worldwide. Forecasts suggest that by 2023 the number of digital Voice assistants will reach around eight billion units – a number higher than the world's population. (https://www.statista.com/statistics/, 2020)

Virtual Assistants

Virtual assistants, an increasingly common feature of many consumer electronics devices, can respond to commands, provide users with information, and assist in the control of interfaced electronics. There are over

110 million virtual assistant users in the United States alone, and the software is usually standard on smartphones and smart speakers. As of 2019, Amazon's Alexa was supported on around 60,000 different smart home devices around the world, providing an excellent example of just how popular the software has become (https://www.statista.com/statistics/, 2020).

"Smart" Everything

Virtual assistants have become a vital component of the smart device industry, absolutely essential to the way that consumers interact with their devices. As the industry grows and its technology becomes more advanced, companies are increasingly searching for bigger and better uses of "smart" technology (https://www.statista.com/statistics/2020). Tech-savvy consumers can now communicate with their connected homes and vehicles in much the same way that they can with their smartphones.

Growth

Shipments of virtual assistants rose 25% year-on-year to 1.1 billion units in 2019. Futuresource estimates the market is on course to exceed 2.5 billion shipments by 2023.

What is causing the shift to Voice? There is a new awareness and comfort level with Voice, specifically with millennial consumers. Speed, efficiency, and convenience are easily optimized with the addition of Voice technologies.

25% of individuals ages 16-24 use Voice search on mobile. (Global Web Index)

Predictions for Voice Assistants

➢ Search behaviors are changing, and more questions are asked—complete, conversational questions. What used to be consumer touch points are being converted into consumer listening points.

➢ Voice assistants will offer increased individualized experiences, even within the same households. Voice differentiation technology continues to improve.

➢ Compatibility and integration with other products will increase interest.

➢ Utilizing Voice push notifications to increase engagement and consumer retention will be a significant focus as new apps are built.

➢ Touch and screened devices will improve, offering more features.

➢ Security will be a focus due to the high volume of users who have questions and issues with trust and privacy policies.

➢ Consumer demographics will continue to grow to include a vast array of populations. According to emarketer, while the millennial generation appears to be the primary consumers of Voice technology, increased usage of Voice-enabled digital assistants can be seen across the multiple generations, including baby boomers.

US Voice-Enabled Digital Assistant Users, by Generation, 2016-2019
millions

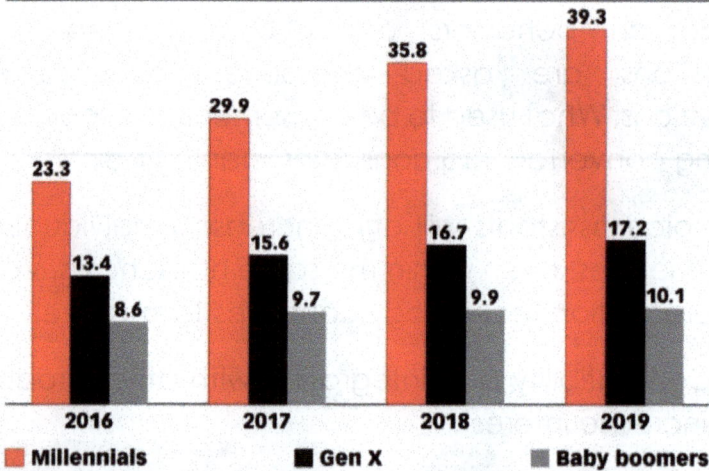

	2016	2017	2018	2019
Millennials	23.3	29.9	35.8	39.3
Gen X	13.4	15.6	16.7	17.2
Baby boomers	8.6	9.7	9.9	10.1

Note: individuals who use voice-enabled digital assistants at least once a month on any device; millennials are individuals born between 1981-2000, Gen X are individuals born between 1965-1980 and baby boomers are individuals born between 1945-1964
Source: eMarketer, April 2017

226458   www.**eMarketer**.com

IS YOUR BUSINESS READY FOR VOICE?

Voice, SEO, and Voice SEO

If you have been practicing best SEO (search engine optimization) practices, you will already be ahead of the SEO for Voice strategy. There are differences in how we must think about SEO for Voice.

When websites began to flood the internet, businesses were told to utilize keywords and optimize for search engines. Due to the overwhelming number of websites, search engines were created to find what you were looking for based on a search query. Then businesses were told a few years later as mobile phones started to hit the market that they needed to optimize for mobile-first. Mobile-first optimization took into consideration things like keywords, location, and page speed.

In 2020 we are going to tell you and explain to you how to optimize your online spaces for Voice search. If you have been consistent with online keywords and mobile-first SEO, it should not take you awfully long to get up to speed with your Voice search optimization. In the past, we focused on keywords, single words, phrases, and small pieces of information, and that, coupled with good content, allowed users to find you and for you to rank higher in the search results. A similar thing happened when we optimized for mobile search. There were a few additions added for mobile, such as page speed, mobile-friendly website, and of course, location.

In **Voice search optimization,** we build upon what is already in place by adding a few new requirements to that which already exists.

1. Keywords now turn into phrases and questions. The fact that we type very differently then we speak must be taken into consideration when discussing Voice search.

For example, instead of typing "pumpkin pie recipe," we are going to say, "How do I make a pumpkin pie?" or "Where can I buy a pumpkin pie?"

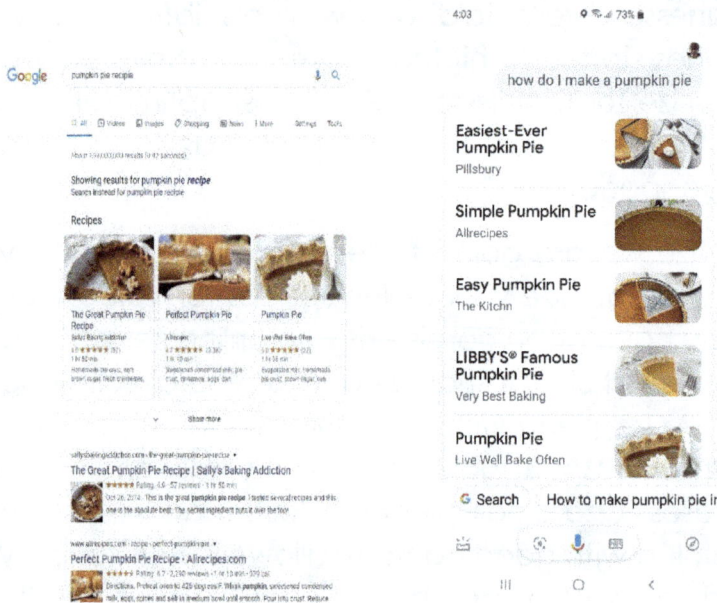

Another example might be, "Let's search for restaurants."

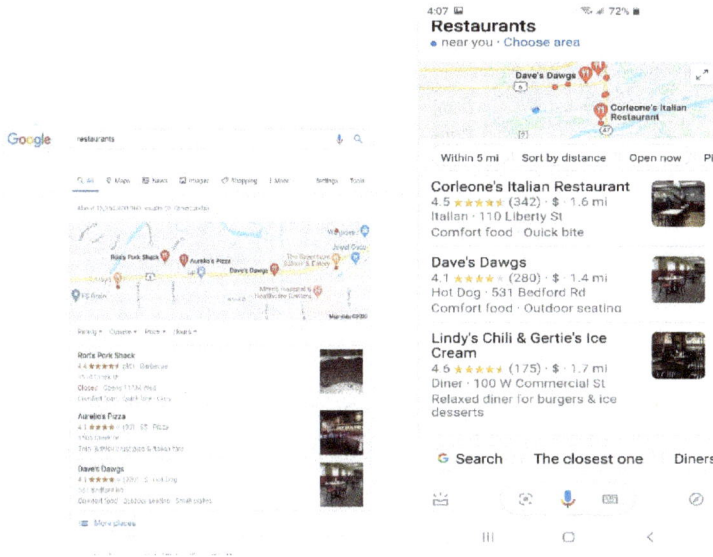

The syntax of our language requires an understanding of long tail phrases, such as **how do I, where, what, and near me,** when optimizing for Voice. The syntax for a written question often times varies from the syntax of a spoken question. In other words, the structure of written language is different than the structure of verbal language.

You should also consider the search results that are delivered to different devices. In a Voicebot.ai study, they break down the different devices and the results shown when searching. (Voicebot.ai, 2019)

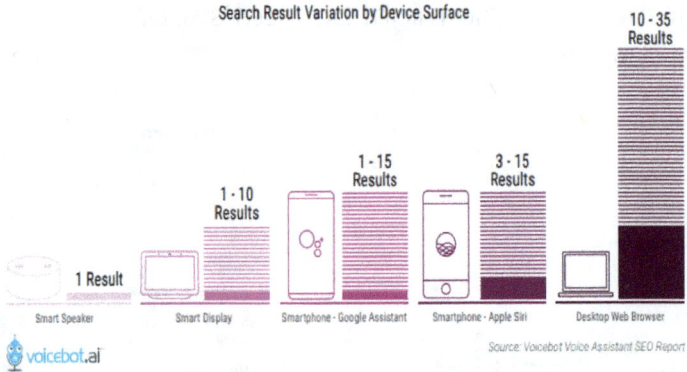

Search Result Variation by Device Surface

Source: Voicebot Voice Assistant SEO Report

In November of 2019, Google updated their search algorithm. BERT (bidirectional encoder representation from transformers.) BERT uses NLP (Natural Language Processing) to consider the context of the words in the query. This change will initially impact 1 in 10 searches but will expand over time.

Voice search really took off on mobile-first. According to the Global Web Index, 25% of individuals ages 16-24 use Voice search on mobile. Over half of smartphone users have been using Voice search since 2015, and usage has increased since then.

Six significant considerations in Voice search

The information that each device absorbs when asked a question comes from a variety of places. Be aware of which sources provide this information. As a business owner or marketer, you want to ensure all information is accurate and useful. A majority of Voice searches are answered via the Rich Answer boxes displayed at the top of the search results. Many of these boxes provide public domain information, such as how many ounces are in a pound, which Google answers using Instant Answers. However, there are other factors to be considered.

Featured Snippets

Featured snippets are displayed on page one of the results, pulled from any website. Google credits brands for these on both regular and Voice search. Featured snippets are more attainable because it is easier to place on page one than position one in the search results. This type of placement may require you only a few minor improvements, rather than a complete overhaul of your existing SEO strategy.

Featured snippets answer questions that help people go places, learn information, buy products, or do something. Perform long-tail keyword research to find these questions and answer them in your content.

Make your content effortless for Google to read and comprehend by framing the question in header tags and answering it quickly. In addition, tabular and bulleted information are also efficient formats for featured snippets.

User Intent

User intent may be defined as the purpose for the search. When creating content, keep the following user intents in mind.

- **I want to know**
- **I want to go**

- **I want to do**

- **I want to buy**

To determine user intents, evaluate the SERP (search engine result pages). Every SERP is unique, even for search queries performed on the same search engine using the same keywords or search queries. The reason is because virtually all search engines customize the experience for their users by presenting results based on many factors beyond their search terms, like the user's physical location, browsing history, and social settings. It may not give you the exact intent, but it will help with your research. Use AdWords to determine the extent of the intent and analyze your analytics.

Longtail Keyword Phrases

Longtail keyword phrases are those 3-4-word phrases used in a search. They are extremely specific. When a consumer employs a specific search phrase, they tend to be looking for exactly what they said in the search. For example, "Find breathable t-shirts." Or "How to lose weight fast."

Page Speed

Page speed gained importance with smartphones and mobile devices and remains significant with Voice technology. Page speed is directly related to search results. It is important to limit items that would slow down your web pages. It is best to test your page speed in a variety of locations.

If a test determines that your site speed is slow, you can try several different adjustments that may increase the speed. Large images can really slow down a website, be sure to use optimized images on your website for faster load time.

✔ Disable unused plugins.

✔ Utilize a preferred hosting provider.

✔ Optimize your images and other visual content for mobile.

People who use Voice search are often in a hurry, and they are busy driving, walking, running, cooking, and doing laundry--typically multitasking. Consumers want answers fast, and they do not want to waste time.

Structured Data

Structured data help search engines understand your content. Structured data assist search engines in creating relevant snippets of information, allowing Voice searches the ability to navigate your site more effectively. For more information on Structured Data, you can visit schema.org.

Local SEO

Local SEO is based on local intent. According to Google, 46% of searches have a local intent.

Focus on the near me intent or location of the device you are using. Google My Business focuses on the GMB map. Therefore, your GMB profile is critical in local SEO marketing for all searches--web-based, mobile, and Voice.

VOICE STRATEGY

As a successful business owner or effective marketing agency, you must add a Voice strategy to your overall marketing strategy. The following are some questions you can ask yourself when building your strategy.

• Who is my audience?

• What problems does my audience have?

• What is my business objective, and can I achieve these objectives by solving the problems my audience is having?

• How am I currently solving the problems of my clients?

• Can I take what I am currently doing and turn this process into a Voice experience?

In addition, consider your audience, their identities, their needs, and their resources.

Assimilating all of this information

There is a process to answering the questions above. Begin by speaking to a sampling of your customers. Ask them questions about the challenges they face, the interests they have, and their lives in general. Then you review email inquiries for purpose and content. More often than not, you will notice that people, in general, ask similar questions regarding a business and the services available. If they have to ask you, perhaps that information can be included on your website and social media platforms.

Then, talk to your customer service representatives as well as your sales force. They are your business's frontline; they typically speak to customers and potential clients every day. They are an excellent source of information about the questions and problems your customer has. Finally, utilize the internet to really listen to what consumers are saying and what their needs are. Once you have garnered all of this information, you are ready to proceed.

Build a robust avatar or your ideal customer. Each business will be different, and you may end up with more than one defined ideal customer. Give this avatar a name, an address, and then plan on how you are going to solve this avatar's problems. Constructing the avatar will allow you to better understand your typical customers. Ask yourself several question to deepen that understanding.

❖ How am I helping Avatar solve their problem?

❖ What services or products do I offer that can help Avatar?

❖ What content can I produce to answer the Avatar's questions?

By improving your customer's experience, you will improve your bottom line. You should set goals for your business, making sure that they are aligned with your customers' needs. The overarching goal should be to create such a positive experience for your customers that they never think about price or discounts, building an experience that keeps them coming back over and over.

A Sample Avatar for a Painting Contractor

Target Persona: Megan
On the Go
Woman between the age of 25-40
Affluent Professionals; high-income ($90K)
Heavy mobile user
Very aware of Social Status
Socially Conscious
Environmentally Conscious
Focused on Career
Owns a Home (Condo, Townhome, Single Family Home)
Lacks Time to work on home
Reads: HGTV, Architectural Digest, Real Simple, Better Home & Garden

Sample Strategy for a Service Business

SAMPLE ONE PAGE MARKETING PLAN
DAWN HARGROVE-AVERY

This is a sample of a marketing plan and items you should consider

Audience	Strategies	Activities
TARGET PERSONA	**MARKETING GOALS**	**MARKET CHANNELS**
who are they (multiple per business)	Increase Sales	In-store
Gender- Age- Employment Status	Increase Pieces	Mobile App
What do they care about?	Increase Tickets	Social / Digital
What are they into?	Grow Audience	TV
Tech Usage, Social causes,	Grow Email List	Voice
hobbies, personality	Increase Followers	Mobile Platforms
CUSTOMER JOURNEY	**KEY STRATEGIES**	**TACTICS & ACTIVITES**
What- Where - How	Build Customer Loyalty	• Inside/outside campaigns: Digital personalization- Voice interaction
What do they do	Improve Customer Experience	• Not all coupons personalized offers based on interactions.
Where do they do it	Expand Mobile Use	• Lifestyle advertising
How will they know?	Expand Voice Use	• Educational content to customers
	Promote Sales & Offers	• Voice interaction orders and conversations.
VALUE PROPOSITION	**PRICING & POSITION**	**MEASURE FOR SUCESS**
What is your companies	Low-Average-Premium	Daily/Weekly/Monthly/Quarterly
promise to the customer?	Quality Level	Average Revenue- Average Traffic
What value do you bring?	Service Level	Loyalty member numbers (mobile/foot/voice
What are diffentiators?	Convenience Level	Transactions
	Social Responsibility	Social Following
	Environmental Responsibility	Web Traffic

PRIORITIES IN VOICE AND VOICE SEO

VOICE SEARCH RESULT CHECKLIST

Fun fact about the state of current directory listings: Only 4% of businesses have complete citations across Google, Yelp, and Bing - the three most important platforms for Voice search readiness. Having accurate and consistent information across all your locations on these 3 directories will provide you with a Voice search readiness score of at least 90% - making your locations Voice search ready. You will also be way ahead of the competition.

Let us dig into the directory checklist, what you should do, how to claim your listing, where you can find your listing and how to better utilize the tools available to you. I have included the actual checklist so you can track each completed task and monitor your progress. The Checklist is available on our website www.VoiceWizards.com. You should revisit this checklist often because as we all know technology changes quickly and new algorithms are released frequently. As a business owner, schedule the time needed to review the progress you have made on the checklist. This reflective practice will allow you to stay ahead of the competition.

Getting Started

Google My Business (GMB)

Step 1: Claim Your Profile

Claim your Google My Business listing. At sign up, you will be asked to add your business information (Address, Opening Hours, Phone Number, Business Name, Website, Zip code, etc.) - Claiming your GMB listing is your #1 Voice optimization priority because Google feeds both Google Assistant and Siri.

Step 2: Embed Map on Your Website

Embed your Google Maps business location on the Contact Us page of your website by identifying your map embed code on your GMB business profile.

Step 3: Create FAQs

One of the lesser-known Voice search optimization strategies is to create your own FAQs in Google Q&A - within your GMB profile. This will allow you to build contextual search value within your profile and rank for long-tail Voice search phrases.

Step 4: Reviews + Engage

Gaining user reviews is crucial when ranking in Voice search for queries where users seek quality ('best,' 'top,' 'highest-rated,' etc.). Customers want you to respond to their reviews. When engaging with customers who leave reviews, you are more likely to influence a negative review when you reply to customer feedback.

Yelp

Step 1: Claim Your Yelp Business Listing

Claim your Yelp listing. At sign up, you will be asked to add your business information (Address, Opening Hours, Phone Number, Business Name, Website, Zip code, etc.). Be sure the information entered is **accurate and consistent with your Google My Business profile information**. As one of the top directories on the internet, Yelp is also important to Voice search optimization. This is because Yelp reviews provide content to Alexa, Siri, and Cortana.

Step 2: Get Reviews + Engage

Being listed on Yelp is important for Voice search 'visibility,' but by engaging with customers whether they leave good reviews or bad, you are more likely to secure return business and change bad reviews to good ones. The better your review score, the better your chances of securing that top spot on local Voice search queries.

Bing

Step 1: Claim your profile

Claim your Bing listing. At sign up, you will be asked to add your business information (Address, Opening Hours, Phone Number, Business Name, Website, Zip code, etc.). You want to make sure that the information entered is accurate and consistent with your Google My Business and Yelp profile information. Bing is the final big player when it comes to your Voice search optimization because it feeds Cortana, Microsoft's virtual assistant.

Step 2: Add images

You should take advantage of the ability to upload images to your listing. Let the people see you!

Step 3: Choose the correct categories

(Think specialties) Categories are tough for some businesses. Choose the correct categories.

Step 4: Be sure information added is the same as GMB

Consistent, standardized information used in Google and Yelp will ensure optimization.

Bing: Quick Start

If you are already on GMB, you can import your information to Bing Places.

Content-Specific Optimization

Optimize short descriptions with a location and neighborhood if it is local. In the long description, you should use a different keyword variation for every 100 words and add the location information. Also, write blogs that cater to commonly asked consumer questions.

Keyword Research

Find online consumer queries that match your business offerings and consider the difference between how people type and how people speak - Voice search queries are likely longer than text and in the form of a question like "how do I change a tire on a car" or a command like "find the nearest Starbucks to me now."

Create a FAQ Page

The typical result of a Voice search is 29 words in length - perfect for a website FAQ page. Identify frequent queries online, create an FAQ page, and write answers that are around 29 words in length to answer them. This is one of the easiest ways to optimize for Voice search.

Write Featured Snippet-worthy content.   40.7%   of Voice search results come from the featured snippets; optimize by integrating keywords concisely.

Embed Long-Tail Keywords in Content

Less than 2% of Voice search results have the exact keyword in the title - insert keywords into the body of the content, correctly answering the question.

Keep Your Message Simple

K.I.S.S. Methodology - Keep it simple for the consumer, think like a consumer, and how they are conducting a Voice search. The average Voice search is conducted at a 14-year-old comprehension level.

Website Specific Optimization

Optimize for Voice search by adding "Near me" in your title tags, meta-description, internal links, and anchor text. Use phrases consumers use to describe the neighborhood around the location. Implement the titles of local institutions and landmarks that are most relevant to the business.

Increase Site Page Speed

The average Voice search result loads in 4.6 seconds - 52% faster than the average page. Find out how to improve your site speed.

Structured Data Optimization

Make your pages relevant and structured with queries in mind. Include Products, Places, People, Organizations, Events, Reviews, How- to Content, etc.

Submit Sitemap to Google

Submit your sitemap from your Google Search Console. From your dashboard, click: Crawl > Sitemaps > Add Test Sitemap.

Prioritize High-quality Pages in the Sitemap. If your sitemap directs bots to low-quality pages, search engines will interpret your site as low quality. Instead, direct bots to pages that include images, videos, and lots of unique content and prompt user engagement through reviews.

Create Actions on Google

There are many reasons why you should have a google app or google action. The most important reason in my opinion is to be everywhere your users/clients/customers are.

Google App

There are a few way's to easily Publish your content using a Google Actions. The easiest way is by using Voice Wizards automatic publishing feature. Simply say **"Hey Google, Open Voice Wizards"** and follow the prompts. Another way is to use the Google account Actions Council.

> ➤ Set your permissions
> ➤ Create your action
> ➤ Add an Action Invocation
> ➤ Add your Action
> ➤ Add Training Phrases for your Intent
> ➤ Add the Action and Parameters
> ➤ Enable Fulfillment
> ➤ Crate your Welcome Intent
> ➤ Test your Action
> ➤ Release your Action

Mobile-first Site Optimization

Optimize your website specifically for mobile. Most Voice search queries come via mobile, so it is crucial to make sure your website is mobile-friendly. You can test your site's mobile optimization and load time here: https://search.google.com/test/mobile-friendly.

MARKETING AND VOICE

Why is voice important and how can you as a business owner or marketer utilize this platform to be where your clients are?

- ✓ **51% of those who shop using Voice, use it to research products.**
- ✓ **2% of these consumers make a purchase directly through Voice.**
- ✓ **17% have used Voice to reorder items.**

How is marketing changing with Voice?

Marketing is changing for several reasons.

- Featured Snippet

- Voice Assistants and Web Traffic

- Voice Results are content driven

- Future Ability to advertise your skills in sponsored ads

- Virtual assistance offers better enabling retentions

- Ability to upsell and cross-sell

- Virtual Assistants are pushing products

- More Conversational Experiences

The Featured snippet area on google is that little box that shows up on google, providing you with instant how to's, definitions, or instructions. This area is now considered position zero on Google search. As of January 2020, be sure to stay informed about the new featured snippet rules that are claiming if your company gets that position for a specific page, you will no longer show up on page 1 of the search results beyond being in the featured snippet section. This does not mean that you will not come up in a search that is tied to a different page on your website. The BERT update may have made this a reality.

Voice Assistants are yielding more web traffic. When you ask a question using one of the Voice assistants, technically it will take you to the website at some point.

Voice search results are providing better content. Business owners are creating better stories. As a business owner or marketer, you can tell a better story using Voice, people can hear your tone and feel the emotions when listening to your voice. Therefore, providing better content and a better experience overall.

In today's market, you can currently only advertise your skill by typical methods such as on your own website, social platforms, and print. With the ability to produce new types of content, and the possibility of new advertising channels, in the future, many of us are highly anticipating the ability to advertise a skill in sponsored ads, in skill advertising, or paid advertising on the various platforms.

Voice assistants are also yielding more insights; access to multiple users in a single provides businesses and marketers with a massive amount of data.

Voice Assistants offer improved enabling retention. This is different from phone apps, and phone apps take up space on your device. Skills and Actions do not, so people tend to leave skills enabled once they have been activated.

As a marketer, you have the ability to inject retention by upselling and cross-selling your products and services. Using the brand message to enhance customer experiences.

Voice Assistants are also able to push products in new ways and enable new experiences such as the Domino's Pizza Easy Order addition or What is my balance at Capital One. Work is recurring between consumers and dry cleaners to manage a closet. For example, if you do not see the pants you are looking for in the closet, you can just ask your dry cleaner if they have your pants.

Voice Assistants are also putting an emphasis on conversations. This makes it easier for someone to multitask--ask questions, order products, set up services while they are cooking, cleaning, and spending time with their families.

Yes, there is more available in Voice assistants, but Alexa and Google currently hold the market share, so this is our focus for now.

What is going to change?

Will Voice replace Facebook or Instagram? What about my website? What type of content do I need? How to create that content?

So many questions have arisen regarding Voice and the changes that will come with it. Let us clarify a few of the frequently asked questions.

First, Voice is not going to replace anything; it is a contemporary technology that is going to be added to the mix that most businesses already have in place. Voice is user-friendly for consumers and businesses alike; it is the natural way humans communicate.

The goal is to meet your customers where they are. Whether your customers are using a phone, computer, smartphone, smart speaker, text messaging, beacon technology, or a simple postcard in the mail, as a business owner you want to get in front of those consumers to provide for their needs.

Building a skill or action does not guarantee new sales. It is like anything you have done from print advertising to digital advertising--websites, apps, mobile. You must advertise it, talk about it, and introduce it to your customers in order to be discovered on the internet. Simply building it does not necessarily mean clients and consumers will come.

Voice technology is advancing rapidly, and you can take advantage of that by staying ahead and understanding the things you can do and control to participate in this new space. There are technologies available in Voice that will make your website Voice readable. You can attach data skills to your website and other online spaces to know where visitors are going. There are many facets of Voice skills: podcasts, flash briefings, customer service, skill purchasing, pickup, and delivery.

Think of all the things you can do today using your Voice. Begin with the most basic, you can send a text message on your mobile device, and you can change the channel on your TV using your Voice.

Voice technology allows versatile flexibility in accomplishing necessary tasks while following the hands-free laws when we are driving. This technology, allowing consumers the freedom to follow these driving laws while simultaneously asking questions, getting directions, locating local businesses, and of course, playing music or listening to books and podcasts on our phones. Currently this technology has become essential in car manufacturing.

Voice technology, smart speakers, and devices now enhance many homes in record numbers, allowing consumers to control various aspects of their homes, look up recipes, listen to music, ask questions, and play games along with listening to audio and podcasts. Our lives have become remarkably busy, and Voice technology is allowing us to multi-task and get things done.

You can throw in a load of laundry while creating a shopping list or listen to your book while cleaning your house. There are so many things you can do by simply using your Voice.

It is easy to see that we are living in a fast-paced world of multitaskers. Voice technology supports this lifestyle helping to navigate consumers so that while you are busy doing one thing you can make a doctor's appointment, book a trip, check your bank balance and in some cases even renew your driver's license all with a simple vocal command.

As a business owner or marketing agency, you must consider the fact that often, consumers and potential customers will be multitasking when using your app. Your skill or action needs to be distinct, useful, and informative.

When creating content, the following questions should be considered:

➢ What type of content should I create for my Voice app?

➢ *How much content do I need?*

➢ *How do I create this content?*

➢ *Where can I find the content I need?*

➢ *How should the content be laid out?*

These are all questions that arise for many small businesses as they develop strategies and skills.

You may already have your first resource for content if you have a website, digital spaces, or brochures about your company. Most of this content is created for mediums that involve us talking at the consumer, not speaking with the consumer. While you will not necessarily be required to create new content; you will need to reframe the existing content so that it works with a Voice platform.

With this in mind, expect that Voice is going to require you to convert your content into conversations, questions, and answers to provide consumers with the impression that they are engaged in authentic human communication.

There are several things to take into consideration when optimizing content for Voice.

> ➢ Understand Voice Search
> ➢ Make Content Searchable
> ➢ Claim Google My Business
> ➢ Focus on Long-Tail keywords
> ➢ Create a FAQ page
> ➢ Use of Schema Markup

Voice search results are typically **29 words**, and the average featured snippet is 45 words in length. A key to effectively creating this content is to answer a question in a short paragraph using approximately 29 words.

USING VOICE FOR AND IN YOUR BUSINESS

Ideas for creating a skill or action for your business

Podcast

Why is a podcast good for your business? There are four main reasons that creating a podcast can be good for your business. 1. Provides an Authoritative Presence

2. Connect with your Audience

3. Simple to create and produce

4. Brand Awareness

Podcasts use a simple audio file to share information through computers and portable music devices along with smartphones. They are easy to create and upload and can provide your customers and clients with information about new products, general company information, and information that relates to your industry.

Flash Briefing

A flash briefing is a short-pre-recorded audio clip, typically less than 10 minutes that is broadcasted over Amazon Alexa devices. Using this new marketing channel allows small businesses to reach a whole new audience by providing tips, industry news, quotes, and professional insights. It is a huge opportunity to get in front of new audiences.

Order a Product

Ordering and reordering products using a Voice assistant is another feature to reduce time calling or going online to place orders.

Schedule a Service

Using calendars connected to Voice Assistance makes it easy to schedule services such as pickup and deliveries, hair appointments, house cleaning services as well as doctor appointments.

Communicate with customers and clients

Think outside of the box and create an experience for your clients, customers, and consumers. Make your Voice experience fun, provide them with a game, or useful advice. These "extras" will keep them coming back.

Skills to utilize in your business and home

Schedule appointments

This Sample is provided by AWS under business productivity:

How to use Alexa to schedule and move meetings.

Scheduling a Meeting with a Contact

Link a supported calendar with Alexa. For more information about how to link your calendar, see Connect Your Calendar to Alexa.

If you've previously linked your Microsoft Office 365 account, you will need to relink your calendar in the Alexa app.

Add work or personal contacts to your Alexa app. For more information about how to add your contacts, see Add and Edit Your Contacts to the Alexa App.

Say '*Alexa, schedule a meeting with John*'.

Let's walk through an example showing how you would use Alexa to schedule a 1-hour meeting with John. Note that if you are a user enrolled with your company's Alexa for Business account, your meeting will default to 30 min to make it easier to find time on busy calendars.

User: *Alexa, schedule a meeting with John.*

Alexa: *You and John are available today from 11:30AM. to 2PM. and tomorrow from 9AM to 11AM. What date and time should I schedule?*

User: *9AM.*

Alexa: *What's the name of the event?*

User: *Finance Review.*

Alexa: *I'll schedule Finance Review on Friday May 11th at 9AM and send an invite to John at john@example.com. OK?*

User: *OK.*

Alexa: *I've added that.*

John then gets an email invite for the event. Alexa suggested up to two free time slots for you and John by looking at your linked work and personal calendars and your contact's linked work calendar.

Now, let us look at other scheduling utterances that you can use.

If you want to schedule a meeting at a specific time, say *"Alexa, schedule a meeting with John today at 3PM."* The Alexa Smart Scheduling Assistant will inform you about the availability for both you and your contact and will book your requested meeting, even if you have a conflict.

If you want to schedule a meeting on a specific day, say *"Alexa, schedule a meeting with John today."* If Alexa can't find available time for you and your contact, Alexa will try to find time where you're busy but the participant is free. Alexa assumes that you'll be willing to make time to meet with the participant because you picked that specific day, signaling that this is an important or urgent meeting. If the participant is busy on that day, then Alexa Smart Scheduling Assistant will look for availability over the next seven days and offer suggestions. (https://aws.amazon.com, 2020)

Moving a Meeting

Link a supported calendar with Alexa. For more information about how to link your calendar, see Connect Your Calendar to Alexa.

Say 'Alexa, move my 9AM meeting' or 'Alexa, move my Finance Review meeting to 4PM tomorrow'.

Let's walk through how to have Alexa move your meeting.

User: Alexa, move my Finance Review meeting to 4PM. tomorrow.

Alexa: OK. I'll move the Finance Review meeting tomorrow from 9AM to 4PM. Just so you know, you have another event named 'Anaya's Soccer Match' at this time. Should I go ahead?

User: Yes.

Alexa: Ok. I'll move the Finance Review meeting to Friday May 11th at 4PM.

The Alexa Smart Scheduling Assistant was able to find the meeting by its title, inform you of a conflict on your personal calendar, and with your approval, move Finance Review to 4PM the next day. Alexa will also send an updated invite email to notify meeting participants of the change.

Send text messages

First of all you need to set up SMS with Alexa, to do this you can launch your Alexa app on your mobile device, go to the communication icon, allow Alexa access to your phone contacts and then Alexa will want to verify your phone number.

Then, to send a message all you will have to do is say "Alexa, send a message", provide the recipient's name when asked, dictate the message, pause when you are done. Alexa will ask if she should send the message and you respond by saying yes.

You can receive a message and when you do your Alexa device will have yellow light, all you have to do is say "Alexa, play my messages."

You can also do this manually using your Alexa app.

Reply to and Read emails

You can also use Alexa to reply to an email or read you your emails. Of course, you will first have to set it up in the app but then all you have to say is.

"Alexa, check my email." Alexa will respond with a summary of new messages from the last 24 hours.

"Alexa, did I get any emails from Bob?" Alexa will prompt you to set up a one-time notification the next time you get a message from that contact.

Improve productivity using these skills

Manage Meetings

In 2020 using Alexa for touch free meetings:

Lifesize Icon 300, Icon 500, and Icon 700 systems with Alexa for Business allow users to simplify their collaboration experience. You can save time by using voice to check into rooms, join meetings, and call contacts. With the Lifesize directory integration, users can easily reach any contact, room system, or meeting in your organization with name-based calling. Room booking utterances, like "Alexa, find me an available room.", allow users to optimize their meeting spaces by finding and booking free rooms, extending room reservations, and automatically releasing rooms that are reserved but not used. Additionally, private skills enable organizations to easily create custom experiences for their employees. For example, "Alexa, what's the guest wifi password?"

With this new solution, company leaders can see helpful usage data and metrics to understand how rooms are being used as employees return to the office. Lifesize Icon also come with the Alexa Built-in certification, which ensures the same security, privacy standards, and features that Alexa offers today. For example, a visual blue bar appears on the screen when Alexa has detected a wake word and audio is streaming to the Alexa cloud. Users can also use a mute icon to disable Alexa at any time. Further, Alexa for Business offers IT and end-user controls to manage and delete the Alexa voice and response history.

Work from Home Productivity (2020)

Information Provided by AWS Alexa for Business

How I set-up my home for remote work and used Alexa for Business work-based features on my Echo devices to boost my work-from-home productivity

Using Alexa to manage my work calendar and email

I started using Alexa to manage my work calendar and email regardless of where I was in my apartment. I would say, "Alexa, when is my next meeting?" or "Alexa, tell me about my workday?" and Alexa would read out the details of my next event (or next few events in case of 'workday'). I would also say "Alexa, schedule a call with Danielle Potter for 3pm tomorrow" and Alexa would create the calendar event and invite the participant using just my voice. These interactions allowed me to stay on top of work when not in range of my screen-based devices, like my laptop and cell phone. To enable these features for my Echo devices, I simply linked my corporate calendar and email accounts in the Alexa app.

Some calendar and email utterances that I now use on a daily basis:

- "Alexa, tell me about my work day"
- "Alexa, what is my work briefing?"
- "Alexa, what are my work updates?"
- "Alexa, prepare me for my workday"
- "Alexa, what's on my calendar?"
- "Alexa, when is my next meeting?"
- "Alexa, schedule a meeting with Danielle Potter for 3pm."
- "Alexa, move/reschedule my meeting"
- "Alexa, cancel my meeting"
- "Alexa, read my emails"
- Actions: "next", "mark as unread", "flag", "delete", "archive", "reply"

Using Alexa to join conference calls, online meetings, and for calling

Alexa also enabled me to join conference calls without opening a dialer or conferencing application on my laptop or cell phone. I would say, "Alexa, join my meeting" and Alexa would check my linked work calendar and connect me to conference calls hands-free. If I needed extra time before a meeting, I would say, "Alexa, I'm running late by 10 minutes" and Alexa would send a short email informing all meeting participants. By using these utterances, I avoided the typical multi-step processes required to join conference calls or to respond to a calendar invitation, freed up my laptop and phone for note taking and messaging, and enjoyed better call quality through my Echo speakers. To start using these meeting and communication features, I linked my corporate calendar in the Alexa app and setup my account for Alexa communications.

Similarly, I began using Alexa for scheduled 1-on-1 meetings and to make quick calls to colleagues. I imported my contact list from my cell phone to use Alexa's native call features like "Alexa, call Matt Otter's phone" and "Alexa, answer the call". For those colleagues with the Alexa app or an Echo Show device, I made Alexa-to-Alexa video calls and transformed the average voice call into a more productive in-person interaction (e.g., "Alexa, call Sean Baker's Alexa device"). As a remote worker, these impromptu video calls were useful in facilitating clear communication with my team. I was able to call any contacts' Alexa devices once they also setup their accounts for Alexa communications.

Some meetings, communication, and calling utterances that I now use on a daily basis:

- "Alexa, join my meeting"
- "Alexa, start the meeting"
- "Alexa, end the meeting"
- "Alexa, I am running late"

Linking Alexa and Office 365

Calendar integration in Alexa for Business

In the AWS Management Console, you can link your calendar system with Alexa for Business. After your calendar system is linked, you can associate your Office 365 resource calendars to the rooms you defined in Alexa for Business.

The calendar integration in Alexa for Business enables your users to join their scheduled meetings, check room availability, and find available meeting rooms by simply asking Alexa. The following diagram and steps describe how Alexa for Business interfaces with the calendar system.

User asks Alexa to join their meeting.

The Alexa Service processes the request, determines the intent, and routes it to the Alexa for Business conferencing service.

The Alexa for Business conferencing service looks up the address of the resource calendar of the room where the request was made.

The Alexa for Business conferencing service connects to Office 365 and reads the upcoming event from the resource calendar.

Alexa for Business determines the dial-in information and prompts the user to confirm the meeting.

After confirming, a call is initiated to connect the video conferencing system to the upcoming meeting. (https://aws.amazon.com, 2020)

5 ESSENTIAL QUESTIONS

Do you really need a skill?

Creating positive customer experiences can help boost your brand image with your clients as well as spread the word about what you have to offer. Also, with the rising boom of AI-platforms and Voice driving assistants, it is becoming essential for businesses to provide a brand experience using Voice.

What will your skill do?

All businesses will have different goals. This is definitely the experimenting and testing phase of Voice, and there is a lot to try and test. Getting things in the proper perspective will help.

Build on your strengths. It is easier to leverage an existing value-proposition than to create a new one. Who is your brand, and how can Voice build from that? What do you stand for?

Instinctively Voice. Consider what your brand has that is or can be uniquely enabled by Voice, e.g., the "hands-free, screenless" moments where only Voice can add value.

Simple value proposition. User interactions incorporating Voice will often be short and direct. This is a great idea for things like status updates, appointment scheduling, and verification, quick and easy purchase, finding quick, straightforward information, tracking packages, tracking an ongoing event.

External triggers. What are the simple triggers that can engage a user to ask your skill something or call your skill into action? This can come from external triggers or daily recurring triggers, what are they?

Whose Voice?

Human Voices are becoming increasingly popular, and the need for brands to have vocal representation on these growing audio-based mediums is expanding (https://www.voices.com, 2019). Synthetic Voice that is readily available on all the platforms is the default. Your brand can stand out by using your brand Voice when creating your skill.

Synthetic Voice: Pros and Cons

Synthetic Voice Pros:

✔ Affordable. Speech synthesizers are inexpensive, in some cases free. Simply enter 'speech synthesizer' into any search engine, and you can easily find a tool that will provide you with text-to-speech capability.

✔ Fast. Amazingly simple, just add your script or text, enter, and a computer will read your lines.

Synthetic Voice Cons:

✔ Unrealistic. The sound quality is not authentic; it is robotic in nature. This does not always lend itself to a positive image for brands or small businesses.

✔ Indistinctive. Many other brands and businesses have probably used this same synthesizer.

AI Voice: Pros and Cons

Artificial intelligence or AI Voice is a type of synthetic Voice, but it operates a bit differently. It differs in that AI Voice uses 'deep learning,' which is a type of artificial intelligence, to turn text into audible human-sounding speech. (https://www.voices.com, 2019)

AI Voice Pros:

✔ Control. You have control over your content and script.

✔ Cost. Programs like this are low cost, so they can be a cost-effective way to process your content.

✔ Immediate results. As soon as you add the words or script, the interpretation of the content is available.

AI Voice Cons:

✔ Ethics. There are some serious ethical issues with robotic Voices masquerading as human Voices and communicating with humans.

✔ Robotic sound quality. They have improved, but these still do not sound quite human.

✔ Soul. The AI Voices have a hard time expressing empathy, humor, and general feelings. This creates a very mechanical conversation devoid of any human essence.

Human Voice-Over: Pros and Cons

Before Synthetic and AI Voice, there was human Voice. The human Voice is unmatched in its ability to convey detailed information that provides connotative meaning beyond the words used. (https://www.voices.com, 2019)

Human Voice-Over Pros:

✔ Authentic. A human Voice will always be a human Voice. The authenticity provides another level of meaning to the communication.

✔ Less legal ramifications. Your Voice belongs to you, and you can be your own brand.

Human Voice-Over Cons:

✔ Expense. Voiceover acting has taken off in the past two years. People have built businesses and careers around Voiceover performance. They assist businesses in creating a brand Voice. This can be costly and might involve licensing agreements.

✔ Notoriety. Reputation is essential, but it is also unpredictable: Sometimes, brands are affected both negatively and positively by the performer that becomes the brand Voice.

What is UX?

What is the User Experience (UX)?

If you want to get technical ISO 9241-210, which describes the ergonomics of human-system interaction, defines user experience as an individual's perceptions and responses that result from use or the anticipated use of a product, service, or system.

The definition of UX can and does fill books. The key point when creating a skill, app, website, or pushing out digital marketing is to create a look, feel, and design as well as the content with the end-user in mind. Think simple, clear, and concise with direct calls to actions.

How do I Market my Skill?

7 things you can do to market your skill

The main goal when marketing the skill is to be sure you have delivered a consistent and engaging skill. Test it to be sure all the bells and whistles work.

1. Optimize Your Skill Name and Invocation Name

Be sure the name you use for your skill is clear and intuitive.

2. Write a Clear, Valuable Description

With so many skills and actions available in the Alexa Skill Store as well as the Google Action Store, make sure your description catches attention.

3. Add an Eye-Catching Skill Icon

Provide clean and eye-catching artwork to create an icon for the different stores. Do not rush; be sure the imagery is clean and clear.

❖ *Keep it simple.*

❖ *Keep text to a minimum.*

❖ *Avoid using photographs. Make it recognizable.*

Now that you have an engaging skill with an optimized store presence, it is time to start talking about it.

4. Send an Email to Your Network

5. Add to Your Email Signature and Update Marketing Icons

6. Feature Your Skill on Your Website

7. Promote Your Skill on Social Media

References

https://aws.amazon.com. (2020, April 20). Retrieved from AWS Amazon: https://aws.amazon.com/blogs/business-productivity/boosting-work-from-home-productivity-with-alexa-for-business/

https://www.statista.com/statistics/. (2020, February 19). *https://www.statista.com/statistics/973815/worldwide-digital-voice-assistant-in-use/*. Retrieved February 19, 2020

https://www.voices.com. (2019, February 12). *https://www.voices.com/blog/guide-synthetic-ai-human-voice-brand/*. Retrieved July 28, 2019

Rosie Murphy. (2018, April 26). *Bright Local*. Retrieved from Bright Ideas Research: https://www.brightlocal.com/research/voice-search-for-local-business-study/

Voicebot.ai. (2019). Retrieved from https://research.voicebot.ai/voice-assistant-seo-report-for-brands/

Closing

In closing, the Voice industry is fairly new and everyone I have had the privilege of meeting in the industry has been so kind and welcoming. The transparency and sharing of information are so refreshing. It is a feeling of openness for the good of the Voice community. From personal experience, I would like to say that the voice industry welcomes newcomers with open arms.

These are fun and exciting times in technological innovations. This is also new technology, and the number of users are growing every day. Predictions for the future of Voice are bright and will continue to evolve.

When I started this book, the year 2020 was a few months out and now as I finish it, we are already halfway through 2020. It has been a remarkably interesting year for everyone, especially small business owners. We have seen an unusual amount of changes in the way small businesses are operating. Voice is a natural contact free way to provide sales, pickup and delivery, customer service and numerus other activities. Including paying bills, banking and so much more.

This is a great time to start working through the process of creating a Voice app for your business. We are extremely excited at Voice Wizard to be able to provide you with a contact free way to get up and running on Alexa and Google Assistant quickly. Let us take a moment to walk you through the process.

You can use Alexa or Google, a speaker or simply your phone:

For Google:

---

1. Hey Google, Open Voice Wizards
2. Add a Business
3. What is the Name of your Business?
4. Say, "Business Name"
5. Can I send you a text message?
6. Yes
7. Provide your phone number
8. Cheers (sound)

For Alexa

---

1. Alexa, Open Voice Wizards
2. Add a Business
3. What is the Name of your Business?
4. Say, "Business Name"
5. Can I send you a text message?
6. Yes
7. Provide your phone number

8. Listen to the Cheers

Once you make it through that process, we will send you videos on what to do next. You can also email us at info@voicewizards.com with all your company information along with your logo and we will get your new skill or action up and running.

Yep it is that simple.

Visit www.VoiceWizards.com for more information.

If you are interested in a voice strategy session you can contact us at Voice Wizards or you can contact our Sister company, Creative Marketing Clinic to assist you with that. We have an entire tribe of people who are waiting to help you with your Digital and Voice Projects.

What are you waiting for? Take a moment to create your first skill, you have nothing to lose it is FREE to simply add your business, and basic contact information to Voice.

Additional information is available on our Website:

Voice apps for your business www.VoiceWizards.com

Planning your skill: Where do I start ?

Like us on Facebook @voicewizards

Follow us on Instagram @voicewizards

Connect with me on LinkedIn Dawn Hargrove-Avery

Visit our Website www.VoiceWizards.com

Add your Business to Voice using Voice Wizards Skill or Action!

www.ingramcontent.com/pod-product-compliance
Lightning Source LLC
Chambersburg PA
CBHW070409200326
41518CB00011B/2133